**BITE SIZE**

# SCIENCE

## 150

### Facts
### You
### Won't
### Believe!

**BITE SIZE**

# SCIENCE
## 150
## Facts
## You
## Won't
## Believe!

**HUGH WESTRUP**
**Pictures by NATE EVANS**

ISBN 0-590-13617-8

12 11 10 9 8 7 6 5 4                    9/9 0 1 2/0

Printed in the U.S.A.          01
First Scholastic printing, January 1997

# TABLE OF CONTENTS

# INTRODUCTION

The world we live in is an amazing place. In writing for a science magazine for students, I have come across some of the most astounding facts.

I hope you enjoy reading these facts and sharing them with your friends and family!

● A crocodile can shoot straight up out of the water and snatch out of the air a bird that is flying overhead.

- The anableps is a fish that has two eyes in each eyeball.

- Many of the snakes in the movie *Raiders of the Lost Ark* weren't really snakes. They were anguids—lizards that don't have legs.

- An owl can turn its head around in almost a complete circle.

- A young golden retriever named Brodie learned to walk again after losing a front and a back leg. Brodie could climb up and down stairs and run as fast as any four-legged dog.

- Shrews are tiny rodents that eat all the time. If a shrew can't get food for even a few hours, it dies.

A male porcupine sings a high-pitched song to attract a female who is ready to mate.

At many zoos, elephants are allowed to amuse themselves by painting pictures on canvases, using their trunks as brushes.

- Some cobras spit venom at the eyes of their enemies. The venom is very painful and can cause permanent blindness.

- The Arctic tern migrates farther than any other bird. It covers up to 22,000 miles per year, flying from the Arctic to the Antarctic and back again.

- Prairie dogs live in large colonies called towns. A prairie-dog town discovered in 1901 had 400 million members and covered 24,000 square miles.

- Small hibernating animals wake up every few weeks to drink, eat, and go to the bathroom.

- Sheep don't swim for fun, but they can dog-paddle their way to safety if they fall into a river or are caught in a flood.

- Snakes smell with their tongues.

- Some fish hold their eggs in their mouth for several weeks until the eggs are ready to hatch.

● Ganymede, one of Jupiter's 16 moons, is the largest moon in the solar system. It is twice as big as the planet Pluto.

● Astronomers used to think that there existed a planet between Mercury and the Sun. They even gave the planet a name: Vulcan.

- Because there is little gravity in outer space, astronauts become an inch or two taller when they leave Earth.

- Europa, another of Jupiter's moons, would make a super skating rink. It is covered by a smooth layer of ice many miles thick.

- Pulsars are stars that spin as fast as 1,000 times per second.

- The planets Jupiter, Saturn, Uranus, and Neptune are not solid planets. They are huge balls of liquid and gas.

- The temperature at the center of the Sun is 15.9 million degrees.

- The largest volcano in the solar system is on Mars. Called Olympus Mons, it is 17 miles high and covers more area than the state of Arizona.

- Pluto takes 247 Earth years to orbit the Sun just once.

- There are about 100 billion stars in our galaxy, the Milky Way.

- Hydrogen is the most abundant element in the universe.

- Jupiter, Saturn, Uranus, and Neptune all have rings around them.

- A black hole is a star that has collapsed. The force of gravity is so great around a black hole that nothing can escape from it. Black holes are invisible because not even light can escape from them.

- Footprints left by astronauts on the Moon will stay there for millions of years because there is no wind on the Moon to blow the footprints away.

- A woman in Alabama was hit by a meteorite—a rock from outer space. It crashed through the roof of her house and hit her on the arm and the hip.

● The Moon is moving slowly away from Earth at the rate of about 2 inches per century.

# MORE AMAZING ANIMALS

- Goats have huge appetites. Hungry goats sometimes climb trees and strip the branches of all their leaves.

- Quahog clams, which can live as long as 220 years, are the longest-lived creatures in the ocean.

- The smallest known dog that ever lived was a Yorkshire terrier that was only 2.5 inches long and weighed just 4 ounces.

- A shrimp that lives in the Great Lakes eats itself. The shrimp eats pieces of shell that it has shed as it has grown bigger.

- Large hares live on the ice and snow in the Canadian Arctic. Called Arctic hares, they bounce around on their hind feet like kangaroos and gather in groups of up to 1,000.

- Toads hop. Frogs leap.

- Many snakes lay eggs, but some give birth to live babies.

- The walking catfish can use its fins and tail to help it crawl on land from one lake to another.

- A flying snake lives in some tropical forests in Asia. It glides from tree to tree.

- A male woodchuck is called a he-chuck. A female woodchuck is called a she-chuck.

Yipe!

- A bat can eat half its weight in insects in one night.

- The tuatara is a lizard with a tail that breaks off easily. If an enemy grabs a tuatara's tail, the tuatara simply sheds the tail and runs away. Later, it grows a new tail.

- Female nine-banded armadillos always give birth to either four male or four female babies.

- The black swallower is a fish that can swallow another fish twice its size.

- The United States has more tornadoes than any other country in the world.

- Rainfalls of animals have happened numerous times in history. People have seen frogs, toads, fish, snakes, beetles, jellyfish, and worms rain from the sky.

- Just before a huge earthquake hit Alaska in 1964, Kodiak bears awoke early from hibernation and ran from their caves.

- Niagara Falls once stopped flowing for two days. Ice in Lake Erie piled up and formed a dam that stopped lake water from flowing into the river that feeds the Falls.

- In the town of Browning, Montana, the temperature dropped 100 degrees—from +44 degrees Fahrenheit to -56 degrees—in a single day.

- A new Hawaiian island is forming on the Pacific Ocean floor. The island is an underwater volcano that is growing bigger and bigger from regular eruptions of lava.

- Killer lakes in the African country of Cameroon occasionally belch clouds of carbon dioxide that smother and kill people and wild animals.

- Lightning occurs more often in Florida than in any other state.

Chickens on a farm in the American Midwest were plucked of all their feathers by strong winds when a tornado touched down on their barnyard.

- Rainbows sometimes appear at night. Called moonbows, nighttime rainbows are totally white in appearance.

- If all the ice at the North and South poles melted, the oceans would rise by as much as 180 feet.

- At the top of Mt. Washington in New Hampshire, the wind blows with hurricane force more than 100 days a year.

- In 1556, an earthquake in China killed 830,000 people.

- Though icebergs float in the salty oceans, most of them are made of freshwater ice.

- Raindrops are not shaped like teardrops. They look more like doughnuts.

- Male and female elephant couples some-times walk trunk in trunk as if holding hands.

- Gorillas are so afraid of water they don't even drink it. Gorillas get all the moisture their bodies need from fruit and plants.

- Giraffes sleep no longer than one hour at a time.

- The tongue of an adult giraffe is 18 inches long and black in color.

18 inch tongue

- An elephant never stops growing during its lifetime. The biggest member of an elephant herd is usually the oldest.

- The touraco is an African bird whose beautiful red color washes away in the rain.

- When hyenas are preparing to go hunting, they make a mad laughing sound.

- A camel can gulp down about 30 gallons of water in just 10 minutes.

- When rhinos aren't getting enough grass in their diet, they sometimes eat the dung of other animals. They eat the dung to get partially digested grass.

- The green mamba is a poisonous African snake that hides on the branch of a tree and drops onto a victim passing below.

- Scientists think the zebra is a black animal with white stripes, not a white animal with black stripes.

- Birds called spur-winged plovers spend much of their time inside the mouths of crocodiles. They help the crocodiles by plucking out and eating leeches that get caught between the crocodile's teeth.

- A group of rhinos is called a crash.

- A rhino's horn is not made of bone. It is made of tightly packed hair.

- The oxpecker is a bird that gets its nourishment by picking ticks out of the hides of zebras, cattle, and rhinos.

- A young elephant feeds on its mother's milk until it is 5 years old.

- When one ostrich yawns, the other members of the ostrich herd also yawn. People watching an ostrich herd yawn usually yawn, too.

- In Africa, there exists a type of chimpanzee called a bonobo. Unlike other chimpanzees, which can behave violently and kill one another, bonobos are peaceful, cooperative creatures.

● The skin of a hippopotamus produces a pink liquid that prevents it from burning under the tropical sun. The liquid also works as a good sunscreen when spread on people.

# GROWING UP—FACTS ABOUT PLANTS

● If you ate nothing but carrots you would turn orange. Carrots contain a substance called carotene that can turn the skin orange if eaten in large enough amounts.

- The sundew plant eats insects. Sticky hairs trap any insect that lands on the plant. Then the hairs wrap around the insect and the plant squirts out a fluid that digests the insect.

- The world's oldest tree is "Methuselah," a bristlecone pine in California. Methuselah is about 4,600 years old.

- A fig tree growing in South Africa had roots that were 400 feet deep.

- The world's tallest grasses are thorny bamboos. They are found in India and grow as high as 120 feet.

- The largest living thing on Earth is "General Sherman," a giant sequoia tree in California. General Sherman is 275 feet tall and weighs about 2,000 tons.

● The part of the broccoli plant that people usually eat is the flower.

Flower

- Broccoli -

- If you touch a Mimosa pudica plant, its tiny leaves will rapidly fold and its branches will fall against its stem.

- Cucumbers, squashes, and tomatoes aren't vegetables. They're fruits.

- Fruits such as peaches and plums that have pits inside are called drupes.

Squash

Peach

● Termites build tall nests out of mud that can be as high as 30 feet.

● Insects breathe through tiny holes along the sides of their bodies.

● There are more kinds of beetles on Earth than any other kind of animal.

- A cockroach can live for several weeks after its head has been cut off.

- After two black widow spiders mate, the female sometimes eats the male.

- A male butterfly can smell a female butterfly located several miles away.

- Earthworms crawl onto the sidewalk during a rainstorm because their holes have filled with water.

- Butterflies are active during the day. Moths are active at night.

- The most destructive insect in the world is the desert locust. A large swarm can eat 20,000 tons of grain and vegetation in one day.

- Insects are eaten by people in many countries of the world. Many insects are good sources of vitamins and minerals.

- A flea can be frozen in ice for a year and return to life when the ice thaws out.

- A flea can jump more than 100 times the length of its body.

- A female tick can suck enough blood in two days to grow up to 200 times her original weight.

- Goliath beetles are so strong that children hitch up toy wagons to them and hold beetle harness races.

- The world's longest insect, the stick insect, is 15 inches long.

- stick insect -

- When a bee finds food, it flies back to its hive and tells the other bees about its discovery by performing a special dance. The dance informs the other bees about the direction and distance of the food from the hive.

★ translation: "...turn left at the cow, then right at the big rock..."

- Mosquitoes bite women more often than they bite men.

- Some ant colonies capture ants from other colonies and force the captives to become slave laborers.

● Boys' hair grows faster than girls' hair.

● People dream for a total of about 2 hours every night.

● A person breathes about 23,000 times a day.

- The human brain weighs only about 3 pounds but uses about 20 percent of the body's energy.

- The strongest muscles in the human body are the masseters. They are located on each side of the mouth and are used for chewing.

- You have hundreds of tiny, eight-legged creatures, called mites, living all over your body.

- Fingernails grow four times faster than toenails.

- The human body has about 5 million hairs on it.

- Most of the dust in a house is made of tiny scales of dead skin shed by the people living in the house.

- Babies don't produce tears when they cry until they are several weeks old.

- A human baby is born with pink lungs. As the baby grows older, its lungs become darker from breathing pollution in the air.

- The human body has about 60,000 miles of blood vessels.

- If you are right-handed, the fingernails on that hand grow faster than the fingernails on your left hand. If you are left-handed, the opposite is true.

- The right lung in humans is larger and heavier than the left lung.

- Some people are allergic to water. Even their own tears cause the skin on their face to develop blisters.

- The human heart beats about 100,000 times in a day.

- The first life-forms on Earth were simple bacteria and algae. They appeared about 3.5 billion years ago.

- All of Earth's continents were once stuck together and formed a huge supercontinent called Pangaea.

- Huge crocodiles, 50 feet long, once lived in the swamps of what is now Texas.

- Dinosaurs often suffered broken bones by accidentally stepping on each other's tails.

- Ferocious creatures called pliosaurs once lived in the ocean. Pliosaurs had huge heads and powerful teeth and jaws, and were able to overpower the biggest sharks.

- Flying reptiles called pterosaurs lived at the same time as the dinosaurs. Pterosaur wings were made of skin, not feathers.

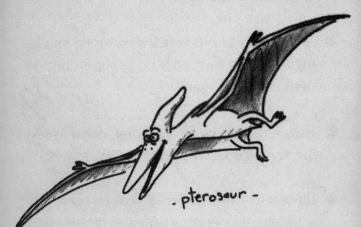

- pterosaur -

- Scientists think the biggest animal that ever flew was a pterosaur called Quetzalcoatlus. It had a wingspan of 40 feet and weighed about 140 pounds.

- The very first bird, Archaeopteryx, descended from a line of small, meat-eating dinosaurs. Some scientists call today's birds living dinosaurs.

- Archaeopteryx -

huh?

- stegosaurus -

● The Stegosaurus, a 2-ton dinosaur, had a brain that was only the size of a walnut and weighed just 2 ounces.

- The Barosaurus was a dinosaur with a 40-foot-long neck. Some scientists think the Barosaurus had eight hearts, which were needed to pump blood all the way up the animal's long neck to its brain.

- North America was once much warmer than it is today. It was so warm that trees grew at the North Pole.

- The heaviest bird that ever lived was Dromornis stirtoni. It weighed more than half a ton.

- North America was once populated by piglike animals called oreodonts. Though their name suggests that they liked cream-filled cookies, oreodonts lived on an all-plant diet.

- A small rodent named Epigaulus lived in North America millions of years ago. On its nose, Epigaulus had two small horns that it might have used for digging burrows.

- A prehistoric elephant that once lived in North America had a mouth shaped like a big shovel. At the end of the shovel were two huge front teeth.

- The largest land mammal that ever lived was a "giraffe rhinocerus" that once inhabited Asia. It weighed 15 tons and was 27 feet long and 18 feet high.

- During the Ice Age, most of North America was covered by glaciers, some of which were almost 2 miles thick.

- North and South America were once home to giant armadillos that were as big as compact-size cars. Their bodies were enclosed in a rigid, bony shell and some had a tail that ended in a large knob bristling with spikes.

# AMAZING AUSTRALIAN ANIMALS

- Koalas are not bears. Like many other mammals that live in Australia, koalas are marsupials—pouched mammals.

- Koalas smell like cough drops because they eat only the leaves of eucalyptus trees.

- Koalas sleep about 22 hours a day.

- A toad that lives in the Australian desert sleeps underground for 11 months a year. During the short rainy season, the toad comes out of the ground to eat, drink, and lay its eggs.

● When a baby kangaroo is born, it is only the size of a bee. A baby kangaroo lives inside its mother's pouch for 33 weeks, feeding on milk and growing bigger.

● Kangaroos are good swimmers.

- Tasmanian devils are marsupials that look like oversized rats. Though they growl and have sharp teeth, Tasmanian devils aren't as fast or as fierce as Taz, the cartoon Tasmanian devil.

- Tasmanian devil -

- Potoroos are small kangaroos the size of rabbits.

- Dingoes are wild Australian dogs. People who catch dingo puppies find that they make good pets.

● Only two types of mammals on Earth lay eggs: the platypus and the spiny anteater. Both are found in Australia.

- spiny anteater -

You're a weird-looking mammal!

- Platypus -

A cassowary is a 5-foot-tall Australian bird that has a bony helmet on its head and dagger-sharp claws on its feet. A cassowary can kill a person with one powerful kick.

- Cassowary -

- The male platypus has sharp spurs on its ankles that hold poison strong enough to kill a dog.

- The bandicoot is a little, pointy-faced marsupial. The bandicoot's pouch opens backward and doesn't fill with dirt as the animal burrows underground.

Hugh Westrup writes for *Current Science*, a children's classroom magazine. He is also the author of several books, including one on prehistoric mammals.